George Frederick Chambers

Weather Facts and Predictions

George Frederick Chambers

Weather Facts and Predictions

ISBN/EAN: 9783337337056

Printed in Europe, USA, Canada, Australia, Japan

Cover: Foto ©berggeist007 / pixelio.de

More available books at **www.hansebooks.com**

WEATHER FACTS

AND

PREDICTIONS.

BY

GEORGE F. CHAMBERS, F.R.A.S.,

OF THE INNER TEMPLE, BARRISTER-AT-LAW.

Author of "A Handbook of Astronomy."

THIRD EDITION.

EAST-BOURNE:

PRINTED FOR THE AUTHOR.

1877.

Price One Shilling.

PREFACE.

THIS Work (first published in 1860; re-issued in 1868, and long out of print) will be found to contain a great variety of statements in the nature of facts, assertions, and predictions concerning the weather. Brevity has been deemed essential to its usefulness, and this consideration has excluded all references to authorities. The strict accuracy of every proposition advanced is not guaranteed; everything must be taken *quantum valeat*; still, it is hoped that nothing palpably unsound will be found herein.

When a rule has any known exceptions, such exceptions are not treated of unless they have some reference to England.

The section devoted to the barometer has been largely drawn from Admiral Fitzroy's voluminous writings, whilst from Buchan's *Meteorology* and Steinmetz's *Weathercasts* some useful information has also been derived.

I shall always be glad to receive corrections and suggestions calculated to increase the usefulness of the book.

G. J. C.

East-Bourne,
    Sussex.

CONTENTS.

—‡—

Weather Facts and Predictions.

———◆———

The Barometer and Pressure.

The barometer is subject to a diurnal variation, comprising two maxima and two minima. The maxima occur within an hour (before or after) of 10 a.m. and 10 p.m. The epochs of minima are, under the same conditions, 4 a.m. and 4 p.m.

The amount of these diurnal variations diminishes from the equator towards either pole. [For the reason that they depend on the influence of the sun's heat on the atmosphere, which is indeed the cause of the diurnal variation at any *one* place.]

There is an annual variation in the height of the barometer, but in England it is of inconsiderable amount.

One of the greatest known changes occurred on September 6th, 1865—a fall of 1·69 inches in 1h. 10m.

In the Northern hemisphere the barometer falls with S.E., S., and S.W. winds: with a S.W. wind it ceases to fall, and at W. begins to rise; it rises with a W., N.W., and N. wind, and with a N.E. wind it ceases to rise, and at E., or towards S.E., begins to fall.

When the wind is on the W. side of the compass changes of barometer *accompany* changes of weather, but with the wind on the

E. side the indications of the barometer *precede* the changes.

Usually a rising or a high barometer indicates less wind or rain, that is, fair weather. A falling or a low barometer more wind or rain, or both, that is, bad weather.

In temperate climates, towards the higher latitudes, the range of the barometer amounts to about 2½ inches, that is to say, it may vary from about 30·8 in. to less than 28·1 in. on extraordinary occasions. The usual range, however, is much less, say from 30·5 in. to 28·75 in. In the tropics the minimum of depression may be put at 27·7 in.

A fall of half a tenth, or still more of a whole tenth, in one hour, is a sure indication that a storm is coming.

If after remaining steady at about 29·9 in. the barometer rises, and the temperature falls, the air at the same time becoming more dry, then N.W., N., or N.E. wind, or less wind may be expected.

If when it falls the temperature rises, and the air becomes more charged with moisture; then wind, rain (or snow) may be expected from the S.E., S., or S.W.

Exceptions to these rules occur when a Northerly wind with wet (rain, snow, or thunderstorm), is impending, previous to which the barometer often *rises* (but only on account of the *direction* of the coming wind), and thus deceives those who from the rising are led to look for fair weather.

The barometer standing at about, say 29·5 in., a rise foretells less wind or a change towards the N. or less wet.

The barometer standing at about, say 29in., the first rising usually precedes high winds from the N.W., N., or N.E., after which, if it still rises, and the temperature falls, improved weather may be looked for. But if the temperature remains high, probably the wind will "back" (shift against the sun's course) and more S. or S.W. wind will follow, especially if the rise has been sudden.

The most violent gales, especially from the N., happen *soon* after the barometer *first* rises from a very low point, or if the wind rises *gradually*, at some time afterwards, although the barometer continues to rise.

A simultaneous rise of pressure and temperature is a sure sign of fine weather. Indications of approaching change of weather are shown by the *movement* of the mercury rather than by its absolute height or depression. Nevertheless, a height of more than 30·0in. is indicative of fine weather, and moderate winds, *except* from E. to N. occasionally, when it *may* blow strongly.

A rapid rise in the barometer indicates unsettled weather, as also does alternate rising and falling.

A gradual rise, or a steady barometer accompanied by dryness, foretells fine weather likely to endure for some time.

A rapid and considerable fall indicates heavy rains with general foul weather.

If fair weather continues for several days, during which the barometer continuously falls, a long succession of foul weather may be expected.

Similarly, if during the continuance of foul

weather, the barometer progressively rises, a succession of fine weather may be looked for.

The greatest depression of the barometer is with gales from about the S.W.; the greatest elevation with wind from about the N.E., or with calmness.

Although the barometer generally falls with a Southerly wind and rises with a Northerly wind, yet the contrary sometimes occurs, in which cases the Southerly wind is dry and the weather fine, and the N. wet and the weather inclement. But sometimes a high barometer (30·2 or so) may be accompanied by heavy rains with a S.W. wind.

When the barometer sinks considerably, high wind, rain, or snow, will follow; the wind will be from the N. if the temperature is, for the season, low; from the S. if the temperature is high.

Sudden falls of the barometer with a W. wind are sometimes followed by violent storms from the N.W., N., or N.E.

If a gale sets in from the E. or S.E., and the wind veers by S., the barometer will continue falling until the wind is near a marked change, when a lull *may* occur; after which the gale will soon be renewed, perhaps suddenly and violently, and the veering of the wind towards the N.W., N., or N.E., will be indicated by a rise in the barometer, accompanied by a fall in the thermometer.

The wind usually veers with the sun (right-handed in N. latitudes, left-handed in S. latitudes); when it does not do so, but backs, more wind or bad weather may be expected.

The barometer sometimes begins to rise

before the conclusion of a gale, sometimes even at its commencement, as the equilibrium begins to be restored.

Though the barometer falls lowest previous to high winds, yet heavy rains often cause a great depression.

Thunder and lightning are *frequently* preceded by a fall in the barometer, but an exception presents itself when the thunderclouds come from the N.E.

Instances of fine weather, notwithstanding the barometer is low, occur from time to time, but they are always preludes to a duration of wind or rain, or both.

After any warm and calm weather a squall or storm with rain may follow ; likewise at any time when the atmosphere is heated much above the usual temperature of the season, and when there is, or recently has been, much electric (or magnetic) disturbance in the atmosphere.

In examining the barometer for the purpose of forecasting the weather, the position of the mercury during previous days or hours should not be neglected, for an indication at any particular moment *may* be affected by causes operating at a distance and not visible to the observer whose barometer feels their effect. Or, as Admiral Fitzroy put it, "There may be heavy rain or violent winds beyond the horizon and the view of an observer, by which his instruments may be affected considerably, though no particular change of weather occurs in his immediate locality."

The longer a change of wind or weather is foretold before it occurs, the longer the presaged weather will last; and, conversely, the shorter the warning the less time will the coming wind or weather continue. In other words—

"Long foretold, long last;
Short notice, soon past."

Sometimes severe weather from the S., *which will not last long*, may cause no great fall, because a northerly wind is impending; and at times the barometer may fall with northerly winds and fine weather apparently against the common rules, because continued southerly wind is impending. Changes thus occurring may mislead, unless the possibility of their occurring is borne in mind.

If the barometer at any place oscillates violently, and the air there remains calm, it is certain there is disturbance going on somewhere in a lateral direction.

At times, in winter, the southerly current prevails over a large area, and the barometer is low and the weather mild. Under these circumstances it may be presumed that severe weather with a high barometer is prevailing somewhere in the neighbourhood. It is possible that this cold and condensed air may force its way into the warm and rarified air in its vicinity, and cause the barometer there to rise rapidly.

Comparing rainy weather with snowy weather, during the course of one revolution of the wind the barometer will fall lower during rain than during snow.

If the barometer *rises* very quickly, it is an indication that the polar and equatorial currents have met and are in conflict. A severe storm is likely to follow, and if the mercury falls as quickly as it rose, the equatorial current has obtained the mastery and the storm is near at hand. In such a case the old barometer phrase "very dry" is wholly delusive.

The barometer is affected by the operation of at least three causes, thus enunciated by Fitzroy :—

1st. The direction of the wind—the N.E. wind tending to raise it most, the S.W. to lower it most, and wind from points of the compass between them, proportionally as they are nearer one or the other extreme point.

N.E. and S.W. may therefore be called the wind's *poles*.

The range, or difference of height shown, due to change of *direction only*, from one of these bearings to the other (supposing strength or force and moisture to remain the same) amounts in these latitudes to about *half an inch* (as read off).

2nd. The amount, taken by itself, of vapour, moisture, wet, rain, or snow, in the wind, or current of air (direction and strength remaining the same) *seems* to cause a change amounting, in an extreme case, to about *half an inch*.

3rd. The strength or force *alone* of wind, from any quarter (moisture and direction being unchanged) is preceded, and foretold, or accompanied by a fall or rise, according as the strength will be greater or less, ranging, in an extreme case, to more than 2 inches.

Hence, supposing the three causes to act *together*, in extreme cases, the height would vary from near thirty-one inches (30·90) to about twenty-seven inches (27·00), which *has happened*, though rarely (even in the tropics).

In general, the three causes act much less strongly, and much less in accord; so that ordinary varieties of weather occur *much* more frequently than extreme changes.

The height of the barometer varies according to the elevation of the place of observation above the level of the sea. Meteorologists having agreed to refer all their barometrical observations to the sea-level as the standard level, it is necessary to add to each barometer reading a certain correction (one-tenth of an inch) for every hundred feet that the barometer read is so elevated above the sea, otherwise barometer readings at different stations would not be mutually comparable.

The Thermometer and Temperature.

In the northern hemisphere the daily range of the temperature is least in winter, augments largely in March and April, reaches a maximum in May or June, continues high during the summer, and diminishes rapidly in October and November to the minimum in winter.

A small daily range, coupled with a high dew point, is thought to be indicative of rain.

The daily range is least in wet climates, and in the tropics and Polar regions, and is greatest in dry climates, and in countries in the temperate zones. [Hence it is less in Ireland than in Scotland; greater in England than in either of these countries, and is greater still on the continent of Europe.]

The daily range in Great Britain in summer is from 12° to 15° in the West and Midland districts, and from 18° to 20° in the South. In the dry climate of Madrid it will sometimes amount to 30°.

The daily range is greater over land than over water, for there is more radiation (disturbance of temperature) from land than from water.

The mean temperature of any given day is the mean of twenty-four hourly observations, but observations at the following hours nearly supply the mean of the day :—

(1.) Between 8 and 9 a.m. and p.m. in the summer.

(2.) Between 9 and 10 a.m. and p.m. in the winter.

(3.) The mean of
$$\left\{\begin{array}{l} \text{4 a.m.} \\ \text{10 a.m.} \\ \text{4 p.m.} \\ \text{10 p.m.} \end{array}\right. \quad \left\{\begin{array}{l} \text{6 a.m.} \\ \text{2 p.m.} \\ \text{10 p.m.} \end{array}\right. \quad \left\{\begin{array}{l} \text{7 a.m.} \\ \text{Noon.} \\ \text{10 p.m.} \end{array}\right.$$

(4.) And generally, the mean of four observations at equal intervals will give the mean for twenty-four hours; as also will the mean of maximum and minimum, without an error, as a rule, of one degree.

When three observations are made daily, the best hours are 9 a.m., 3 p.m., and 9 p.m. The observation at 3 p.m., being near the time when the temperature is at its highest, is of great value in reference to the climate of a locality, as well as in reference to other considerations of more strictly scientific interest.

The mean of observations at hours of the same number (or name) a.m. and p.m. do not differ much from the true mean of the day ascertained under circumstances of precision. This is especially true of 9 a.m. and p.m., 10 ditto, 3 ditto, and 4 ditto.

Where the daily range is small, that is to say in the tropics, and in temperate regions in winter, the maximum temperature occurs at about 1.30 p.m.; but in temperate regions in summer not until between 2.30 and 3.30 p.m.

In winter, and at night in dry calm clear weather, the air is warmer at some height above the ground than it is at the surface. [This explains why fog, which is vapour condensed by chilled air, is so frequently visible in low-lying places, whilst neighbouring eminences are clear.] In such cases the upper rooms are warmer than those nearer the ground—a consideration for invalids.

Houses most protected against severe weather are those on a gentle acclivity, a little above the plain or valley from which it rises, and which have a southern aspect, with trees on the rising ground in the rear.

Comparing Great Britain with places on the continent or in America having the same latitude, the comparative mildness of the former is due to the influence of the Gulf Stream.

It is an undoubted fact that the mean temperature of Great Britain is higher than it was some centuries ago. This is due to the drainage of land generally and to the reclamation of Waste lands. Glaisher considers that the mean temperature of the year at Greenwich has risen 2° in the last 100 years and

that the increase of temperature is especially
marked for the months of November,
December and January.

The winter temperature of Great Britain is
so distributed that for invalids a journey
southward is of little benefit, unless directed
at the same time towards the W.; and as
the W. temperatures from Wales to Shetland
are uniform and equal to those of Sussex, it
is only the south-western counties that present
maximum temperatures. These are $4°$ in
excess of the West of Scotland and Sussex,
and $6°$ in excess of the East of Scotland and
England.

The S. W. of Ireland may be compared
with the corresponding part of England.

The greater the range of temperature, comparing summer with winter, the greater, as a
rule, is the death-rate. Hence the greater
mortality of England compared with
Scotland.

The mean annual temperature falls on an
average about $1°$ for every increase of 300 feet
in the height above the sea.

When the atmosphere is highly charged
with vapour, the escape of heat by radiation
is obstructed, and the temperature falls but
little during the night; but when the quantity
of vapour is small, radiation is less impeded,
and the temperature falls rapidly.

Similarly in the day-time; the prevalence
of vapour obstructs the passage of the solar
rays, and the temperature rises slowly; but
the absence of vapour causes a rapid rise.

The comparative absence of aqueous
vapours in mountainous districts, facilitating

radiation both solar and terrestrial, causes the heat of the sun to fall upon tourists with scorching effect.

When the air is what we call sultry, it is saturated with moisture, and evaporation from our bodies proceeds sluggishly. Hence the well-known oppressive sensation often felt in the summer months, especially in July. Under those circumstances there will be very little, if any, difference between the readings of the wet and dry bulb thermometers.

At any time when the air has been for a while much heated above the usual temperature of the season, a sudden squall, with or without rain, will often come on without much warning.

A period of excessive cold is often followed by destructive gales of wind.

In the northern hemisphere the thermometer rises with E., S.E., and S. winds. With a S.W. wind it ceases to rise, and begins to fall: it falls with W., N.W., and N. winds; and with a N.E. wind it ceases to fall and begins to rise.

The thermometer (shaded from the sun and freely exposed to the air) when much *higher* between 8 and 9 a.m. than the *average*, indicates southerly or westerly wind (equatorial); but when considerably lower, northerly (Polar) currents.

The average temperatures at Greenwich in the shade and exposed in air are *nearly the mean temperatures of each* 24 *hours,* taking the year through, around London. Making an allowance for the difference between the mean annual temperature of Greenwich and of any other particular place the average temperature for the *middle* of each month at such place may be obtained approximately from the following table :—

	Deg.		Deg.
January	37	July	62
February	39	August	61
March	41	September	57
April	46	October	50
May	53	November	43
June	59	December	39

—and proportionally for periods intermediate between the 15th of each month.

The Hygrometer and Moisture.

Since wind drives away saturated air, and so causes dry air to take its place, evaporation is greater in windy than in calm weather.

A rise in the dew-point from morning to noon will be followed by rain : a fall by fine weather.

The evening dew-point generally determines the mean temperature of the night. If, therefore, this be ascertained, the approach of low temperature or of a frost at night may be seen and provided against.

A high evening dew-point indicates, if the dry bulb does not fall much, that the next day will probably be warm, but a high evening dew-point with a chilly air and a S.W. wind is rather a presage of rain.

Damp air is a much better conductor of heat than dry air, consequently it feels colder than dry air of the same temperature, because it conducts away more rapidly the heat from our bodies.

The difference between the dry and wet

bulb thermometer will in England sometimes
amount to 18°, and frequently be from 9° to
12°. This sort of thing will occur between
April and September. During the Winter
months the difference will be restricted to
narrower limits, say from 4° to 9°.

The Temperature of the Dew-point is sometimes as much as 30° below the temperature
of the air : and between April and September
especially, is frequently 20°. During the
Winter months the difference is much smaller,
but is often between 6° and 15°.

When in summer a hot day is not followed
by a dew, rain may be looked for.

A profuse dew is a very sure sign of fine
weather.

Winds.

Two principal currents blow over the
northern hemisphere of the earth. The
Equatorial northwards *to* the Pole, and the
Polar, southward *from* the Pole.

The Equatorial current is warm ; the Polar
current cold. [Because winds bring with
them the temperatures of the regions which
they have passed over.]

Winds coming from the sea do not cause
such variations in temperature as those coming
from a continent. [Because marine temperatures are more uniform than continental
ones.]

Moist winds blowing from the Ocean are
accompanied by a mild temperature in winter,
and by a cool temperature in summer.
[Because air loaded with vapour obstructs
both solar and terrestrial radiation.]

Similarly, dry winds from a continent bring cold in winter and heat in summer.

The Equatorial current becomes a more moist wind as it proceeds N. [Because it loses heat, and therefore approaches nearer the point of saturation, or, as some have it, because it is blowing from regions of much moisture to regions where moisture is less.] The Polar current becomes a more dry wind as it proceeds S. [Because it gains heat, and therefore recedes from the point of saturation, or, as some have it, because it is blowing from regions deficient in moisture to regions of much moisture.]

Hence in England the S.W.* wind is particularly moist [because it is both an oceanic and an Equatorial wind], and the N.E.* wind is particularly dry. [Because it is both a Polar and a continental current.]

Western borders of continents in the N. temperate zone where the prevailing wind is S.W. enjoy a comparatively high temperature in winter. [Because they are protected from extreme cold by the warmth brought by the said wind from the ocean in their proximity, and they are further protected by their moist atmosphere and clouded skies.]

But in the interior of the continent it is otherwise. [Because the S.W. wind getting colder and drier as it advances, the soil is exposed to the full effects of radiation during the long winter nights, and as the ground is for the most part covered with snow, little

---

* Why currents N. and S. at their origin become N.E. and S.W. by the time they arrive in Britain, is a question of physical geography. The fact is due to the Earth's rotation.

heat can ascend from the soil below to counteract the cold on the surface, and so the temperature falls considerably.]

It is much hotter in the interior of such continents in summer than at the Western borders. [Because the land being warmer than the ocean at this season, the wind becomes warmer as it traverses the land, and the superjacent air being drier, the rays of the sun act with an intensity which is always more or less excessive.]

Wind blows from regions where the barometer is high to where it is low, and with a force proportioned to the difference of the pressures, and places *between* very high and very low pressures feel most the violence of the resulting storm, and *not* those where the pressure is absolutely the lowest.

If the wind being N. passes to N.E. we get clear weather; the air is dry, the barometer high, and in winter a considerable degree of cold follows. If the wind gets on to the E. the barometer will fall, and the sky become more or less overcast. Snow and S. wind may then be expected. If the barometer falls rapidly, the snow will turn to rain, and a thaw set in if the wind veers farther, through S.E. and S. to S.W.

If a N.E. wind be accompanied in winter by a clear sky, with haze near the horizon, and the barometer be high and rising, or at least stationary, and the wind does not increase in force but tends to change in the direction of E. and S.E. the weather will probably continue settled for some time.

One of the surest signs of the breaking up

of a severe frost is the setting in of a N.E. wind if it be accompanied by a green or yellowish green sky, and the break up is all the more certain if the sky becomes gradually overcast and the wind backs from N.E. to N. and N.W.

Of S.E. wind there are two distinct kinds, one with a low barometer accompanied by warmth and moisture, ending in rainy or stormy weather. The other with a high barometer rising or stationary, and accompanied by dry weather and a clear sky which may be expected to last some time.

Prolonged absence of wind is favourable to the prevalence of epidemics; but during continued windy weather, no disease arising out of local causes, such as deficient water or drainage, can be expected to make much progress.

In England windy weather is most common in December and January; then in February and November. The calmest months are August and September.

In England S.-Westerly winds prevail most; hence it may be inferred why the W. end is so often the fashionable quarter of large towns, because smoke, &c., is driven from the W. towards the E. Such a current not only carries away from the W. its own smoke, but keeps away altogether smoke contributed by Eastern districts.

The following represents the prevalence of wind in England on an average of years :—

N.	41	S.E.	20	W.	38
N.E.	48	S.	34	N.W.	24
E.	23	S.W.	104	Calm	33

c

When the wind becomes N.E. and two days pass without rain, and on the third day the wind does *not* veer to the S. nor does rain fall, the wind will probably continue N.E. for eight or nine days all fair and then veer to the S.

The wind is usually more strong in veering from N. to W. by S., than in veering from N. to S. by E.

Change of wind "with the sun" (veering N., E., S., W.) is a general indication of fine weather, but "backing" (N., W., S., E.) presages rain or wind, or both combined.

Changes of wind in Ireland, Wales, and Cornwall usually precede changes in the midland and Eastern counties of England by one and a half to two days.

If in unsettled weather the wind veers from S.W. towards N.W. at sunset, an improvement in the weather may be looked for.

Strong winds are more persistent and uniform than light winds. That is to say, if a strong wind is blowing at any one place, a similar wind also strong will prevail over a considerable tract of country, but if the movement of the air is feeble, very different winds will be registered at places not very far apart.

In our latitudes a Northerly current becomes more Easterly the longer it lasts, and, therefore, a N.E. wind is a N. wind which has come from higher latitudes than a wind which reaches us as a N. wind; and similarly, a S.W. wind is a S. wind which has come from lower latitudes than our S. wind.

In Europe the coldest point of the compass is about N.E. in winter and N.W. in summer, and accordingly the warmest winter point is S.W. and the warmest summer point S.E.

When a calm is succeeded by a forward motion of the wind (E., S., W., N.) it seldom " backs," but if it back it will generally return to the point whence it started, before performing a complete circuit.

If the wind be S. for two or three days, it may be succeeded unexpectedly by a northerly breeze; but a northerly wind will not be followed by a southerly one till after the intervention of a period of E. wind.

As a main characteristic, a N. wind is *cold;* and an E. wind *dry;* a S. wind *warm, but rarely wet;* and a W. wind *generally rainy.*

If a S.S.E. wind commences to blow gently, it will freshen gradually. If the sky becomes overcast, the wind will rise and may shift towards the S.W.

If the wind veers, it attains its maximum intensity between W.S.W. and W.N.W., and it never remains long in N.W. and N.

A shifting of the wind backwards may be expected to presage atmospheric disturbance of greater or less intensity.

In spring, if the wind shifts through W. to N. we may expect the weather to clear up suddenly and night frosts to set in, even though the thermometer at a little height above the ground may not fall to 32°.

The only winds which can preserve their directions unaltered in passing over a large tract of country are due E. and due W. winds

[The directions of all others are influenced by the Earth's rotation.]

The backing of the wind when it extends beyond S. or E. indicates cyclones.

Shiftings of limited extent, as N.W. to S.W., or E.N.E., to N.N.E., are often only the return of the vane to its original position owing to the Equatorial or Polar current, as the case may be, regaining its supremacy.

When a calm succeeds a storm, the pressure is unusually low, consequently the foul air imprisoned in the mineral of coal pits escapes more readily into the air, accompanied by a buzzing sound, which miners regard as prognosticating a storm or heavy rain. Accordingly, it is when the barometer is low that explosions of fire-damp are most common.

Storms are invariably vorticose, that is to say, travel across a region in curved paths which do not return on themselves.

Squalls occurring during storms are thought to indicate the approaching cessation of the storm.

Just prior to a tempest the atmosphere is unusually still. [Because its great rarefaction (low-barometer) enfeebles its ability to transmit sounds.]

After a gale from S.E. or S.W. to N.W. a lull of a day or two may follow with symptoms of a continuance of bad weather, then a rapid *backing* of the wind through W. and S. to S.E. After this, another gale may spring up which may be even more violent than the former one.

If the wind howls, or veers about much, rain will follow.

If (towards sunset especially) the sky clears on any part of the horizon the wind will shortly blow from the quarter cleared.

When the sea gets rough on a flood tide it is a sign of more wind coming [S. coast of England.]

Ozone.

When ozone is largely present in the air, it is accompanied by diminished pressure, increasing temperature and humidity, and a prevalence of S.W. or Equatorial winds. But when it is present only to a limited extent, the pressure is increasing, the temperature and humidity decreasing, and the prevailing wind is N.E. or Polar in character.

Ozone is more abundant on the sea coast than inland, in the West of Great Britain than in the East, in elevated than in low situations, in rural districts than in towns, and on the windward than on the leeward side of towns. Its amount seems to increase and decrease with the electricity, and it is almost wanting in places where there is much decaying vegetable or animal matter.

Ozone is most abundant in May, least so in November.

Clouds.

An attentive consideration of the changing tints of the evening sky after stormy weather furnishes valuable aid in forecasting the weather. If the yellow tint becomes a sickly green, more rainy and stormy weather may be expected, but if it deepens into orange or red the atmosphere is becoming drier, and fine weather will follow.

Small thin clouds high up in the E. sky before sunrise, and which soon disappear, are sure prognostics of fine weather.

A green or yellowish green sky is one of the surest signs of rain in summer and of snow in winter.

A red or yellow sky in the morning betokens wind and unsettled weather.

If the clouds move rapidly, or possess in the N.W., a leaden hue, rain may be looked for.

If at sunset the clouds begin to break up and disappear, and have their edges tinged with red or golden yellow, the weather is likely to remain fine and settled.

CIRRUS is an important prognostic of stormy weather, but small groups of regularly formed cirrus scattered over the sky often accompany settled fair weather. Horizontal sheets of cirrus which descend quickly, and pass into cirro-stratus, indicate, unmistakeably, wet weather.

When streaks of cirrus run quite across the sky in the direction in which a light wind is blowing, the wind will probably soon blow hard but in one uniform direction. There will be none of the variable squally weather which usually accompanies storms.

When fine threads of cirrus appear as if swept back at one end by a breeze prevailing in the regions in which they lie, the wind on the earth's surface may be expected to veer round to that point if then at some other point. If the direction so foreshadowed be S.W. (whence the storms of Europe come), wind and rain will follow, and no matter how settled the weather may seem to be, a storm more or less severe is advancing, and will present itself within thirty or forty hours. When the form *seems* past, and the sky is clear, should a few fine cirrus clouds be seen brushed back at their E. extremities, the storm in all probability is really past and fair weather setting in [because the dry Polar current is asserting its supremacy overhead]; but if the cirrus continues to prevail in all directions, interlaced in the sky, a second storm is approaching.

If cirrus forms during fine weather with a falling barometer, a change is sure to occur.

Cirrus is especially associated with easterly winds.

A mass of cirrus with the fibres pointing upwards denote rain, but with the fibres downwards, dry though possibly windy weather. When cirrus lies from W. to E. a storm is imminent.

CIRRO-CUMULUS ("mackerel sky") occurs frequently in summer in connection with dry summery weather. When enduring, cirro-cumuli clouds indicate a continuance of such weather and a rise in the barometer.

CUMULO-STRATUS immediately precedes the fall of rain or snow, according to the season of the year.

CIRRO-STRATUS is especially a precursor of storms. Its greater or less abundance and permanence afford a clue to the probable nearness or remoteness of the storm that is impending.

Since it possesses great extent and evenness of texture with slight depth, it is the cloud in which halos and such optical atmospheric phenomena present themselves, and therefore it is that such phenomena are to be regarded as presaging foul weather.

CUMULUS of moderate height and size, with well-defined curved outlines, and visible only during the heat of the day, especially if they come up with the wind, indicate a continuance of fine weather.

But when they increase with great rapidity, sink down into the lower parts of the atmosphere, and do not disappear towards evening, rain may be looked for. If loose fleecy patches of cloud begin to be thrown out as it were from their surfaces, especially if they move against the wind, the rain is at hand.

If cumuli diminish in size towards evening, they betoken fine weather, but if they increase foul weather. Large masses of cumulus following rain often precede squalls of hail or rain.

NIMBUS (rain cloud): when this is seen approaching, and cirri are noticed to be shooting out from the top in all directions, as the cirri are more numerous, so the rainfall will be more copious.

STRATUS clouds forming at sunset and breaking up as the sun rises in the heavens, and soon disappearing altogether, indicate a continuance of very severe weather.

When clouds are seen drifting about aloft, the air on the ground being still, or nearly so, wind is approaching, and the clouds indicate the direction from which it will come.

Clouds drifting about at sunset, from whatever quarter, betoken rough weather.

Dusky clouds, or clouds of the hue of tarnished silver are a sign of hail. If traces of blue are visible the hail will be small, if traces of yellow it will be large.

It was the opinion of Sir J. Herschel that "anvil-shaped" clouds are likely to be speedily followed by a gale of wind.

The most cloudy countries are those where the wind is most variable, as Britain; the least cloudy countries are those in which the wind is least variable, as Egypt.

If small clouds increase, expect rain.

If large clouds decrease, expect fine weather.

Soft flimy clouds indicate fine weather, with light breezes.

Hard-edged clouds, wind.

A dark, gloomy blue sky indicates wind.

A light, bright blue sky fine weather.

The softer and more silky clouds look, the less wind (but perhaps more rain) may there be expected; but the harder, more tufted, more ragged, the stronger the expected wind will be.

Fragments of clouds ("scud") driving across heavy masses of cloud presage wind and rain, but if scud alone is flitting about, the indication may be merely of wind.

Clouds up aloft at great elevations, crossing the sky in a direction different from that indicated on the surface as the direction of the wind, foretell a change of wind towards their direction.

In general it may be said that light, delicate, quiet tints or colours, with clouds of soft, ill-defined outline, indicate and accompany fine weather; but gaudy or unusual hues, with clouds of hard outline, foretell rain and wind.

Mists.

Mists often appear sooner on parts of hills covered with trees than elsewhere. This happens especially when the mist begins to form after midday, because then the temperature of the trees is lower than that of the grassy slopes. Similarly, mists often linger longest over forests, probably on account of the comparative cold of forest air, arising from the large evaporating surface presented by the leaves of the constituent trees, which, be it remembered, are all the while enshrouded in mist.

Cloudy mists forming or hanging on heights indicate, if they endure, increase, or descend, that wind and rain are coming, but if they rise or disperse fine weather may be looked for.

A morning mist which breaks up into soft looking cumuli clouds betokens a fine day.

Frequent mists foretell rain.

A gloomy mist especially presages rain. [The gloominess is owing to the presence of black clouds over head, which will supply the rain.]

Mists in autumn are often followed by wet; in spring seldom.

If in summer towards dusk, a mist is seen to rise from a stream or a meadow, the next day will be warm.

If a mist appear before sunrise about the time of full moon, fair weather for some days may be expected.

Mists are usually observed with barometrical extremes—very high or very low pressure. [A high barometer testifies to a period of calm,

during which vapour has been able to accumulate ; a low barometer testifies to rarefaction
of the air, which is consequently incapable of
sustaining much vapour in suspension.]

Fogs.

Fogs do not occur in windy weather. They
are driven away when a breeze springs up,
unless dissipated by other causes.

If in winter a cold and a warm current meet,
and the latter (a Southerly one) is overcome
by the former (a Northerly one) the barometer
will rise to a high point at places near the
line of contact, and a dense fog will appear.
This fog often disappears suddenly and then
reappears, and perhaps such alternations may
occur several times: the alternate predomi-
nance of the two antagonistic currents is
indicated by this. If great cold ensue it will
be a proof that the Northerly (or Polar) current
has eventually gained the mastery.

Dew and Hoar Frost.

Dew is the aqueous vapour of the air
deposited on surfaces cooled by radiation.
The quantity depends on the degree of the
cold and on the radiating and conducting
power of the surfaces. Furs, wool, silk,
cotton, vegetable substances, &c., being good
conductors (relatively) will be much bedewed.
Glass, mould, sand, gravel, &c., being bad
conductors, will be little bedewed. By a
benificent arrangement therefore of the Creator
dew falls most copiously on the objects which
most require its refreshing influence.* It is

* Buchan.

not deposited in cloudy weather, because the clouds obstruct the escape of heat into space by radiation, nor in windy wheather because wind constantly changes the air in contact with the ground, and thus prevents its temperature from falling sufficiently low. When the temperature falls below freezing the dew becomes converted into *hoar frost*.

As dew is not formed during the prevalence of clouds or wind it serves as an indication of fine weather.

Hoar frosts on three successive mornings in early spring or autumn betoken rain, but in April or May are generally followed by dry weather.

Rain.

More rain falls on land than at sea, especially in hilly or mountainous countries, and so the temperature will be raised by the latent heat thus given out. [For this reason the northern hemisphere, as containing more land, is warmer than the southern hemisphere.

The drainage of agricultural land has been proved to raise the mean annual temperature, and so such operations improve the produce of the crops (to the detriment of the wells).

Places having a considerable rainfall are characterised by a low mean pressure. More showers of rain happen between 2 and 3 p.m. than during any other hour. Rain is rare between midnight and 1 a.m.

In weather which is showery rather than steadily wet, when between the intervals of sunshine a cloud appears in the W., passes over the spectator, and as it passes pours down a considerable quantity of rain, if the barometer be watched from the time of the cloud's appearance in the W., to its appearance in the E., it will be observed to fall a little and then to recover its original level, the fall being quite a local one.

A steady continuance of rain is usually preceded by fair weather.

If after rain seems imminent several days of fine weather occur, it is certain that the condition of the air has undergone change; that the aqueous vapour has been driven off before it had time to condense. A change of wind will accompany such a general change.

If much rain from the S. is followed by fine weather for a week, the wind remaining S., a prolonged absence of rain may be expected.

Rain usually comes from the W. A clear sunset is therefore a proof that no rain is imminent from that quarter nor probably very near from any quarter.

In winter rain with a W. wind and a rising barometer turns to snow.

Sudden rains do not last long, but when the sky thickens gradually and the heavenly bodies grow more and more dim, rain of some duration may be looked for.

If it begins to rain before sunrise the rain may cease by about noon; but if the rain commences after the sun has risen it is likely to last all day, unless it should be preceded by a rainbow.

A heavy shower after the commencement of a gale of wind indicates an approaching calm.

More rain falls in summer than in winter, and most in the autumn.

More rain falls by night than by day. [Because the cold at night condenses and cools the air, and thus diminishes its capacity for holding moisture in suspension.]

The amount of moisture in the atmosphere is greatest near the equator, and diminishes towards the poles.

The zone of greatest moisture follows the sun across the equator to the North or to the South as the sun's declination changes.

The regions of greatest heat are also the regions of greatest rainfall. More rain falls in the Northern hemisphere than in the Southern.

As to the rainfall in Great Britain—

More rain falls on the Western than on the Eastern coasts, in the ratio of 2, 3, or 4 to 1.

Localities having a small annual fall have most rain in the summer, but at wet stations, winter is the season of most rain.

In all except mountainous districts, the amount of rain increases about $2\frac{1}{2}$ per cent. for every increase of 100 feet in elevation above the sea level.

The wettest place in the British Isles is the Stye Head Pass, one mile S. of Seathwaite in Borrowdale, where the average annual fall is 165 inches.

The driest district in England is that around Lincoln, where the average annual fall is only 20 inches.

Comparing the E. with the W. of England the average falls over the whole area, neglecting extremes, are 25 inches in the former, and 40 inches in the latter.

The fall in the driest year will be one-third below the average; that in the wettest year one-third beyond. Therefore the fall in the wettest will be double that in the driest year.

And, therefore, an excess or defect of say 20 per cent. entitles us to call any particular year a "wet" or "dry" one as the case may be.

July, August, and October are the wettest months at most lowland stations, but December, January, and February, are the wettest in the mountainous districts.

The rainfall is more evenly distributed through the year at Western coast stations than it is at Eastern or dry stations, or it may be put in this way:—The heaviest of heavy falls, say in twenty-four hours, at a wet station, will not amount to 6 per cent. of the annual total, whilst at a dry station a heavy fall may amount to 10, 12, or 14 per cent. of the whole annual quantity.

The question whether the rainfall of England is increasing, diminishing, or remaining stationary, must be answered by saying that observations extending over more than 140 years show, if anything, a slight increase.

The average rainfall for all England is $31\frac{1}{4}$ inches; for near London 25 inches.

In the British Isles, rain falls on an average on 183 days during the year. On 90 of those days the fall will be less than one-tenth, yielding about 4 inches; and on 15 days will exceed $\frac{1}{2}$ an inch, yielding about 11 inches.

It will happen at least once during a term of years, at some period or other, that there will be a specially heavy fall of between 3 and 4 inches, and it may be said that as far as is known no part of Great Britain is free from the chances of such a visitation.

As long as the diameter of the guage exceeds 3 inches it makes hardly any difference in the rain taken whether the diameter of the guage be 4 inches or 24 inches.

When the Greenwich rainfall of the first seven months of the year has been large (say 14 inches or more) the mean temperature of the following winter (December — February inclusive) will be in excess of or about the average, unless the mean of the intervening period of August—October inclusive, has been remarkably cold, in which case the latter part of the succeeding winter (say February) will present some marked extremes of cold.—(Brumham.)

Rain may be expected when the sky assumes an almost colourless appearance in the direction of the Wind, especially if lines of dark or muddy cirro-strati lie above and about the horizon.

Thunderstorms.

Thunderstorms are most frequent in the tropics and diminish in frequency towards the poles. They are more frequent in summer than in winter and in mountainous countries than in plains, during the day than during the night, and after midday than before.

Just prior to the bursting of the storm the air is exceptionally warm and stifling, and this characteristic is especially noticeable in winter and at night.

After the storm is past a great fall occurs in the thermometer.

Thunderstorms coming up with an E. wind while the barometer is falling do not cool the air ; it remains sultry, and another thunderstorm may be looked for as at least likely to occur in the neighbourhood. Not till the wind gets round towards the W., and the barometer begins to rise, will the temperature of the air fall.

If several thunderstorms come on in succession from the W., each storm has usually a more northerly drift than the one which preceded it.

Thunderstorms in spring lie at a low level and do not last long ; they are usually followed by a period of cool weather.

The general direction of a thunderstorm is either from E. to W., or from N. to S. ; not often is its direction oblique.

Thunder occurs commonly when the wind is S. ; very rarely when it is N.

In summer or early autumn, if after the wind has been S. for two or three days, the air becomes very sultry, and numerous clouds with white summits and blackish bases present themselves, thunder and rain are imminent ; if two independent masses of such clouds are seen approaching on different sides, a storm is very near.

Lightning.

There is more lightning in summer and autumn than in winter and spring.

That form of lightning known as "silent"

or "heat" lightning (unattended by thunder)
is in Scotland a prognostic of unsettled weather,
it being considered the forerunner of an approaching storm.

Hailstorms.

Hailstorms are very local in their character,
seldom occur during night or in winter, but
most frequently in summer and during the
hottest part of the day.

Snow.

Snowflakes vary in size from 1-14th inch to
1 inch. They are largest when the temperature
is near 32°, and smallest at very low temperatures. The crystals of the same fall of snow
are generally similar to each other, but one set
of crystals will differ a good deal from another.

When snow descends in very large flakes
with a southerly wind, there will soon be a thaw.

Usually, snow with an E. wind and a falling
barometer turns to rain, the wind becoming
more southerly.

If after severe cold it begins to snow, and
the wind veers from E. to S.E., and the barometer
falls and the cold becomes less intense, still the
thermometer may remain below 32°. In such
a case, when the wind reaches the S., the snow
does *not* turn to rain, and if the southerly current is displaced the snow will continue almost
or quite uninterruptedly.

Snow with a W. wind brings more cold,
though, perhaps, after an interval. [Because
snow is more usual with W. than with E.
winds, and W. winds are apt to work round

and become N. winds, whilst E. winds commonly work round towards the warmer part of the compass.]

Snow can never fall when the temperature is very low. What sometimes appear to be snowflakes with a very low thermometer are rather spiculæ of ice dropped from a stratum of clouds belonging to a warmer current, which happens to be over-head at a great distance. These spiculæ passing in their fall through very dry air cannot increase in size, and therefore cannot assume the form of flakes, because flakes are aggregations of frozen material.

Sleet.

Sleet falls chiefly in winter and spring, and is very rarely an accompaniment of storms.

The Rainbow.

If the predominating hue is green, more rain may be expected; if red, wind and rain.

"A rainbow in the morning is the shepherd's warning ;
A rainbow at night is the shepherd's delight."

A morning rainbow is considered to prognosticate wet, stormy, weather. [Because at that time of day moisture ought to be diminishing; but the presence of rain shows that that is not the case; nay, more, that the moisture is really augmenting.]

An evening rainbow is a presage of fine weather. [Because the conditions under which a rainbow can then appear is the passing away of the rain-cloud to E.; in other words, a clearing up in the W., and that at a time of day when the temperature is on the decline and condensation would naturally be expected.]

A rainbow after much wet is a prelude to fine weather, especially if it breaks up all at once.

A rainbow in the morning is thought to presage much fine small rain; one at noon, heavy rain in torrents.

Two or three rainbows on the same day may accompany weather on the whole fine, but very much rain will follow in a few days.

Coronæ and Halos.

Coronæ* betoken rain when their diameters contract, and fine weather when they increase.

Halos round the Sun or the Moon are nearly sure to be followed by rainy and unsettled weather.

The Aurora Borealis.

This meteor is generally followed within 24 hours by rain and gales of wind, or general unsettled weather.

During the continuance of an auroral display ozone abounds.

Sunrise and Sunset.

If the clouds present at sunrise break up or work off to the W. as the Sun's elevation increases, a fine day will follow.

If the Sun rises distorted in form, showers may be looked for in *summer*, but settled weather in winter.

---

* Often seen and not to be confounded with Halos.

When the first indications of daylight are seen above a bank of clouds, it is called a "high dawn." When the day breaks on or near the horizon, the first streaks of light being low down, it is called a "low dawn." A high dawn presages wind; a low dawn fair weather. If the Sun rises red, with blackish beams, in a haze, rain may be expected; if the western sky is red, wind.

> "If red the Sun begins his race,
> Be sure that rain will fall apace."

A grey or pale yellow sunset is a sign of rain; a bright yellow sunset betokens wind.

> "Evening red and morning gray
> Set the traveller on his way;
> Evening gray and morning red
> Bring down rain upon his head."

If the Sun sets behind a straight skirting of cloud, wind may certainly be expected, and from the point of setting.

If the Sun sets in a clear sky, with its outline sharp and of a deep salmon colour, in summer a very fine, and probably hot, day will succeed, but in winter a very sharp frost.

A red or orange sunset indicates that the following day will be fine, especially if there is much dew; but it may be windy.

If the Sun sets in thick clouds, and the E. horizon is red, purple, or coppery, rain may be expected; if in a white haze, so that its disc can scarcely be discerned, wind.

The Twilight.

If after sunset the W. sky acquires a purple cast with a haziness in the horizon, the following day will be fine. But should the predominating colour be pale yellow extending high towards the zenith, there will be a change of weather.

If the twilight is unusually protracted, though the atmosphere seems very clear the higher regions are charged with moisture which will soon be precipitated in the form of rain.

If a cloudy day clears up at night, rain may follow. [The clouds are dissipated by sinking into a warm lower stratum of air, but the moisture remains, and it is only a question of time when it is precipitated.]

The Moon.

If the Moon is pale and the cusps blunt, rain is indicated; but fine weather if it is clear and its outline sharp.

When the Moon is near its full (on either side), as it rises in the heavens, clouds frequently break up and eventually disperse as night wears on.

A "young Moon with the old Moon in its arms" is a sign of rain, being equivalent in physical effect to the clearness of distant hills, which clearness always betokens impending rain.

The Stars.

If the stars are clear and twinkle brightly, fine weather in summer and frost in winter are indicated.

If the stars are dull and large, and are devoid of rays (twinkling imperceptible), rain may be expected soon.

When the stars twinkle excessively, wind is indicated.

Seasonal Predictions.

Cold weather often prevails about March 8, April 11, April 25, and the second or third week in May, and warm weather about the middle of August, the first week in November, and the first week in December. Hence, when at these times the weather begins to turn warm or cold, a continuance of such weather may be expected for a few days.

Another authority gives the following table:—

COLD PERIODS.		WARM PERIODS.	
February	7—10	July	12—15
April	11—14	August	12—15
May	9—14	December	3— 9
June	29—34		
August	6—11		
November	6—12		

The cold weather of the beginning of November is accompanied by an unusual prevalence of N. and N.W. winds and calms; by a diminution of the invisible vapour in the atmosphere; by an increase in the rain-fall, both as to its amount and its frequency; and by frequent fogs.

There is some connection between cold weather and the Easter full moon. It is very rare for the fortnight after Easter to pass without snow falling somewhere in England.— (*E. B. Denison.*)

If after a prevalence of S.W. wind a N.E. wind should set in, it is highly probable that winds more or less easterly will prevail for some time. If the season be winter, frost, and perhaps snow, may be expected; if summer, the weather will become dry, warm, and bracing, without being sultry or oppressive.

If easterly winds have largely predominated in autumn, and south-westerly winds have begun to prevail towards the end of November, or at the beginning of December, exceptionally mild weather with storms of wind and rain will follow till about Christmas. Indeed this sort of weather occurs almost every year, and its beginning is popularly known as St. Martin's summer.

If easterly winds predominate in spring largely above the average, the summer is likely to be characterised by S.W. winds with much rain and moisture, and little sunshine. If easterly winds nearly fail in spring, they are likely to prevail in summer, and the summer will be characterised by clear skies, and dry warm bracing weather, with much sunshine.

There is strong reason to believe that whatever be the prevalent direction of the wind during a period of about 10 days at either equinox, wind of much the same direction will on the whole prevail during the succeeding three or four months, or even up to the next equinox.

The first fortnight in January, (or days from January 6—20,) usually embraces the coldest period of the year.

Good harvests depend (*meteorologically*) on a warm sunny July and August. The prevalence of sunny weather in June ripens crops prematurely, and the grain is deficient in size.

Crops of wheat are best after hot dry summers; beans after cool moist summer; oats and barley after summers of intermediate character.

In general, a spring characterised by rainy mornings will be followed by an autumn with rainy evenings, and *vice versâ*.

If much rain falls between February 10 and March 10, then the spring and summer quarters are likely to be more or less wet.

In general, a rainy winter is followed by a dry autumn in the following year, and a dry spring by a rainy autumn.

If the latter end of October and the early part of November be for the most part warm and rainy, then January and February will be frosty and cold, except when the previous summer has been very dry.

If there has been low temperature, still more if snow, in October or November, then January and February will be mild.

On an average of years, July, especially the second fortnight, is very wet.

A turning point in the winter season often occurs about January 18 : cold weather either sets in markedly or ceases.

The severest winters are those which begin about January 6, having been preceded by a mild December.

Forests and Trees.

It seems well established that when the oak is in leaf before the ash, a dry summer is to follow.

Forests and trees increase the rainfall, and

diminish evaporation from the soil ; make the days cooler and the nights warmer. [Because changes of temperature take place slowly in trees, but rapidly in air, and also because trees obstruct nocturnal radiation, and, when the locality is a slope with trees at the top, they obstruct the descent of currents of cold air.]

Soils.

The surfaces of loam and clay soils do not heat as do the surfaces of sandy soils, because the former soils being the better conductors, the heat does not accumulate on the surface, but is propagated downwards ; and for the like reason the surfaces of rocks are cooler than either.

Arid regions are commonly deficient in vegetation, and by consequence are deficient in rain-fall.

The Sea.

The sea and deep lakes moderate nocturnal winter temperatures. [Owing to the high specific heat of water, its temperature falls more slowly than that of land ; and when radiation cools the surface layers, they sink to the bottom, and are replaced by other layers which are of higher temperatures. A constant change is therefore going on, and the result is that the superjacent air is very little chilled on calm nights.]

Mountains.

Mountains deprive of their moisture winds which cross them, and thus cause cold winters

and hot summers in places to the leeward, as
compared with places to the windward, [by
more fully exposing them to both solar and
terrestrial radiation.]

Miscellaneous.

After fine weather, the first signs of a change
are usually light. streaks, curls, wisps, or
mottled patches of white distant clouds, which
gradually consolidate into a general overcast-
ing of the sky.

Usually, the higher and more distant such
clouds seem to be, the more gradual but *general*
the coming change of weather will prove.

When the outlines of distant hills are very
clearly seen, and waterfalls and railway whistles
are very distinctly heard, rain is indicated.
[N. and E., that is dry winds, are suddenly
followed by a moist wind, the vapours of which
condensing on particles of dust floating about
in the air makes them heavy, and they sink to
the ground.]

When, in winter or spring, during rough,
sleety, or rainy weather, fires seem to crackle
and throw out more heat, the weather will
probably soon clear up, with frost or frosty air
following.

A gray, dull, morning is nearly sure to turn
to a fine, clear, day.

When frogs dive below the surface of ponds
&c., out of the way, rain is certainly impending.

www.ingramcontent.com/pod-product-compliance
Lightning Source LLC
Chambersburg PA
CBHW022106210326
41519CB00056B/1641